Adapted Tests

AF861888

PROPERTY OF
WENTZVILLE R-IV
SCHOOL DISTRICT

PEARSON
Prentice Hall

Boston, Massachusetts
Upper Saddle River, New Jersey

Copyright © by Pearson Education, Inc., publishing as Pearson Prentice Hall, Boston, Massachusetts 02116. All rights reserved. Printed in the United States of America. This publication is protected by copyright, and permission should be obtained from the publisher prior to any prohibited reproduction, storage in a retrieval system, or transmission in any form or by any means, electronic, mechanical, photocopying, recording, or likewise. The publisher hereby grants permission to reproduce these pages, in part or in whole, for classroom use only, the number not to exceed the number of students in each class. Notice of copyright must appear on all copies. For information regarding permission(s), write to: Rights and Permissions Department, One Lake Street, Upper Saddle River, New Jersey 07458.

Pearson Prentice Hall™ is a trademark of Pearson Education, Inc.
Pearson® is a registered trademark of Pearson plc.
Prentice Hall® is a registered trademark of Pearson Education, Inc.

PEARSON
Prentice Hall

ISBN 0-13-166584-7
7 8 9 10 11 10 09

Astronomy

© Pearson Education, Inc., publishing as Pearson Prentice Hall. All rights reserved.

Earth, Moon, and Sun

Multiple Choice

Write the letter of the correct answer on the line at the left.

____ **1.** The path that Earth follows around the sun is called its
- **a.** rotation.
- **b.** axis.
- **c.** orbit.

____ **2.** Earth's spinning on its axis is called its
- **a.** revolution.
- **b.** rotation.
- **c.** orbit.

____ **3.** When the south end of Earth's axis is tilted toward the sun,
- **a.** it is summer in the Northern Hemisphere.
- **b.** it is winter in the Southern Hemisphere.
- **c.** it is winter in the Northern Hemisphere.

____ **4.** A push or pull is a
- **a.** force.
- **b.** mass.
- **c.** phase.

____ **5.** The tendency of an object to resist a change in motion is
- **a.** orbit.
- **b.** inertia.
- **c.** gravity.

____ **6.** What causes the phases of the moon, eclipses, and tides?
- **a.** the tilt of Earth on its axis
- **b.** the changing relative positions of the moon, Earth, and sun
- **c.** the masses of objects and the distance between them

____ **7.** The darkest part of the moon's shadow is the
- **a.** axis.
- **b.** penumbra.
- **c.** umbra.

____ **8.** The moon formed when
- **a.** Earth began orbiting the sun.
- **b.** a large object crashed into Earth.
- **c.** the first solar eclipse occurred.

© Pearson Education, Inc., publishing as Pearson Prentice Hall. All rights reserved.

____ 9. Maria on the moon are
  a. regions formed from huge lava flows.
  b. regions with many craters.
  c. large seas filled with salt water.

____ 10. The moon's surface gets very hot in direct sunlight because
  a. the moon has no liquid water.
  b. the moon tilts on its axis.
  c. the moon has no atmosphere.

Completion

Read each word in the box. In each sentence below, fill in the correct word or words. Not all words will be used.

solstice	weight	full moon	neap tide	equinox	crater

11. A day on which the sun is overhead at either 23.5 degrees north or south is called a(an) ______________________.
12. The force of gravity on an object is its ______________________.
13. A tide with the smallest difference between low and high tides is called a(an) ______________________.
14. A lunar eclipse occurs only when the moon is in the ______________________ phase.
15. A large round pit on the moon is called a(an) ______________________.

True or False

If a statement is true, write true. *If it is false, write* false.

________ 16. In one year, Earth completes 365 revolutions.

________ 17. Earth has seasons because its axis is tilted as it revolves around the sun.

________ 18. A solar eclipse occurs during a full moon when Earth is directly between the moon and the sun.

________ 19. When the moon is in Earth's umbra, you see a total lunar eclipse.

________ 20. The strength of gravity depends on the masses of the objects and the distance between them.

© Pearson Education, Inc., publishing as Pearson Prentice Hall. All rights reserved.

Using Science Skills

The picture below shows a person standing on Earth at points A, B, C, and D. The picture shows the sun and four positions of the moon as it orbits Earth. Notice that in each position, the moon is sunlit on only one side. Use the picture to answer questions 21, 22, and 23.

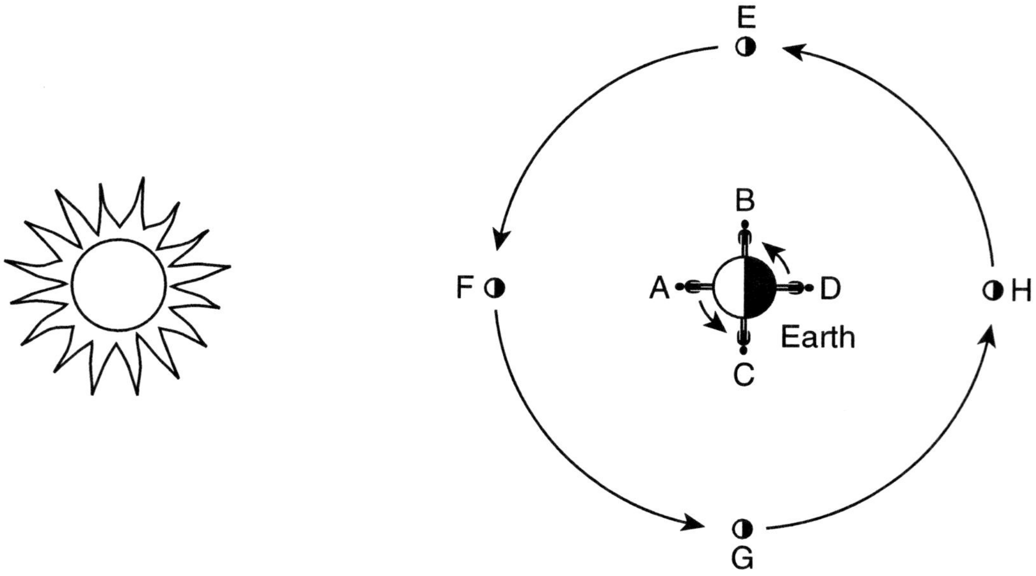

____ **21. Interpreting Diagrams** On which place on Earth would a person experience nighttime?

a. A
b. C
c. D

____ **22. Interpreting Diagrams** Which position of the moon shows a new moon?

a. F
b. H
c. G

____ **23. Applying Concepts** In which position of the moon could a total lunar eclipse occur?

a. F
b. G
c. H

© Pearson Education, Inc., publishing as Pearson Prentice Hall. All rights reserved.

Using Science Skills

The picture below shows the sun, Earth, and the moon revolving around Earth. Use the picture to answer questions 24 and 25.

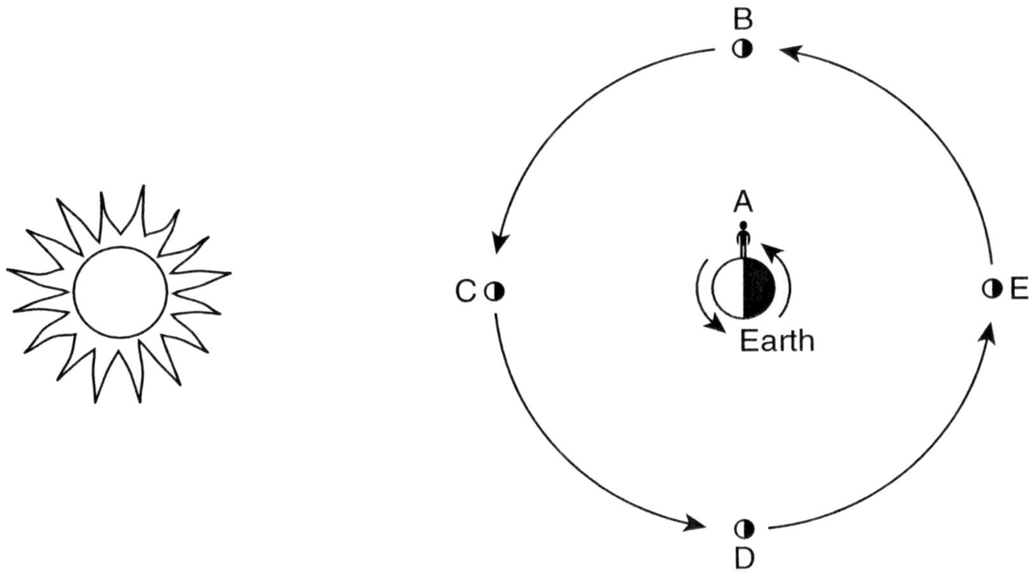

____ **24. Interpreting Diagrams** In which position of the moon would Earth be experiencing spring tides?

a. B
b. C
c. D

____ **25. Applying Concepts** The person standing on Earth at point A sees the moon in its third quarter. At that time, what tides would the person also see?

a. neap tides
b. highest high tides
c. spring tides

© Pearson Education, Inc., publishing as Pearson Prentice Hall. All rights reserved.

Exploring Space

Multiple Choice

Write the letter of the correct answer on the line at the left.

____ **1.** The force that pushes a rocket forward is called
  a. velocity.
  b. thrust.
  c. gravity.

____ **2.** Rockets were first developed in
  a. the United States.
  b. China.
  c. the Soviet Union.

____ **3.** What is the main advantage of a multistage rocket?
  a. The total weight of the rocket is reduced as the rocket rises.
  b. The rocket can be reused after returning to Earth.
  c. It requires less fuel to overcome gravity when it is launched.

____ **4.** The space race was between
  a. the United States and the Soviet Union.
  b. the Soviet Union and Great Britain.
  c. the United States and China.

____ **5.** Who was the first person in space?
  a. John Glenn
  b. Yuri Gagarin
  c. Alan Shepard

____ **6.** Humans first landed on the moon in
  a. 1957.
  b. 1961.
  c. 1969.

____ **7.** The goal of the Apollo program was to
  a. put the first American in space.
  b. land an astronaut on the moon.
  c. prove that the moon was solid.

© Pearson Education, Inc., publishing as Pearson Prentice Hall. All rights reserved.

____ **8.** Which kind of spacecraft can carry satellites into orbit?
  a. space probe
  b. space shuttle
  c. space station

____ **9.** A large artificial satellite on which people can live is called a(an)
  a. space station.
  b. observation satellite.
  c. space probe.

____ **10.** An astronaut orbiting Earth experiences microgravity because
  a. gravity is not pulling on the astronaut.
  b. the astronaut is falling around Earth.
  c. the mass of the astronaut increases in space.

Completion

Read each word in the box. In each sentence below, fill in the correct word or words. Not all words will be used.

rovers	vacuum	satellite
space spinoffs	space probe	velocity

11. The speed of a rocket in a given direction is its ______________________.

12. An object that revolves around another object in space is a ______________________.

13. Some space probes have small robots called ______________________ that collect samples from another planet.

14. Space has extreme temperatures because it is nearly a ______________________.

15. Joysticks and memory metals are examples of ______________________.

© Pearson Education, Inc., publishing as Pearson Prentice Hall. All rights reserved.

True or False

If a statement is true, write true. *If it is false, write* false.

________ **16.** Escape velocity is the velocity a rocket must reach to fly beyond a planet's gravitational pull.

________ **17.** In a solid-fuel rocket, both the oxygen and the fuel are in liquid form.

________ **18.** The first American to orbit Earth was Alan Shepard.

________ **19.** A space probe is a spacecraft that has no human crew.

________ **20.** Observation satellites collect data using remote sensing.

Using Science Skills

The picture below shows Earth and a spacecraft that is orbiting Earth. The arrows from the spacecraft represent two different factors that keep the spacecraft in orbit. Use the picture to answer questions 21, 22, and 23.

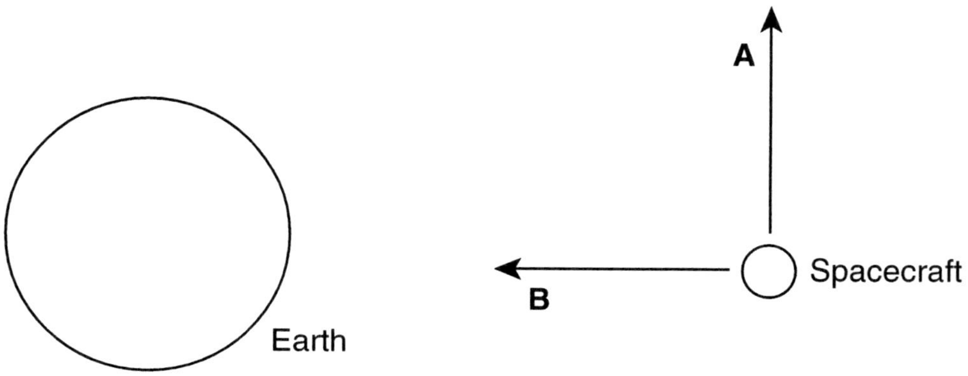

____ **21. Interpreting Diagrams** What does the line labeled **A** represent?

a. gravity
b. inertia
c. friction

____ **22. Interpreting Diagrams** What force does the line labeled **B** represent?

a. gravity
b. inertia
c. friction

© Pearson Education, Inc., publishing as Pearson Prentice Hall. All rights reserved.

____ **23. Predicting** If Earth suddenly disappeared, what path would the spacecraft take?

a. It would move in the direction of the arrow **A**.

b. It would move in the direction of the arrow **B**.

c. It would move in the opposite direction of arrow **A**.

Using Science Skills

Each picture below shows the position of a satellite that is in a geosynchronous orbit around Earth. Use the pictures to answer questions 24 and 25.

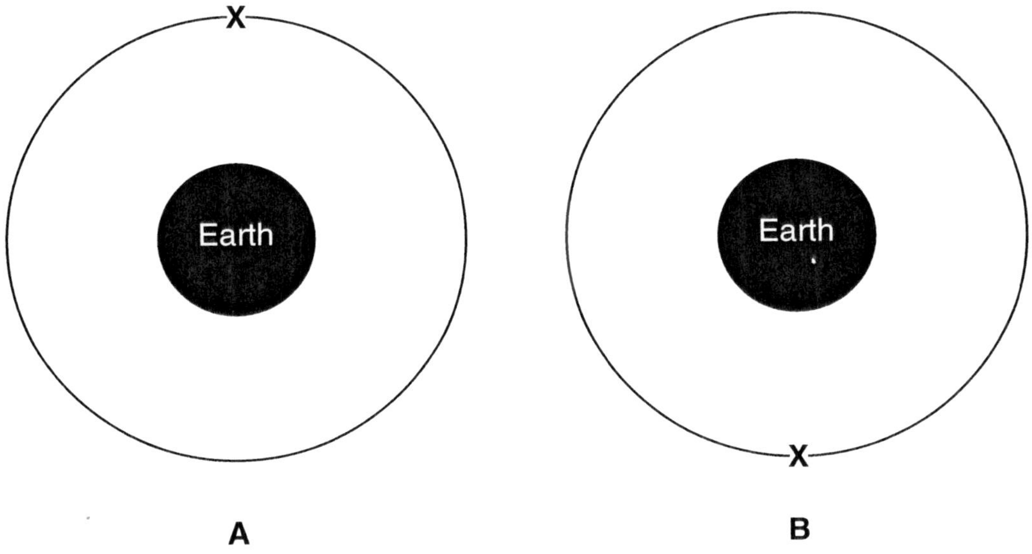

____ **24. Interpreting Diagrams** The X in Figure A shows the position of the satellite at 6:00 A.M. Later that same day, the satellite is in the position shown by the X in Figure B. What time is it in Figure B?

a. 6:00 A.M.

b. 9:00 A.M.

c. 6:00 P.M.

____ **25. Applying Concepts** Where will the satellite be the next morning at 6:00 A.M.?

a. It will be at the X in Figure B.

b. It will be at the X in Figure A.

c. It will be midway between the X in Figure A and the X in Figure B.

© Pearson Education, Inc., publishing as Pearson Prentice Hall. All rights reserved.

The Solar System

Multiple Choice

Write the letter of the correct answer on the line at the left.

____ **1.** The scientist who first used a telescope to support the heliocentric model was
- **a.** Galileo.
- **b.** Ptolomy.
- **c.** Newton.

____ **2.** What are loops of gas on the sun that link different parts of sunspot regions together?
- **a.** solar flares
- **b.** solar winds
- **c.** prominences

____ **3.** What process provides the sun with energy?
- **a.** solar winds
- **b.** nuclear fusion
- **c.** greenhouse effect

____ **4.** The "red planet" is
- **a.** Venus.
- **b.** Uranus.
- **c.** Mars.

____ **5.** Which planet is the smallest terrestrial planet?
- **a.** Mercury
- **b.** Earth
- **c.** Mars

____ **6.** The planet with the most spectacular rings of any planet is
- **a.** Neptune.
- **b.** Saturn.
- **c.** Jupiter.

____ **7.** The inner planets all have
- **a.** a set of rings.
- **b.** many moons.
- **c.** rocky surfaces.

© Pearson Education, Inc., publishing as Pearson Prentice Hall. All rights reserved.

____ **8.** A loose collection of ice, dust, and small rocky particles that orbits the sun is a(an)

a. asteroid.
b. meteoroid.
c. comet.

____ **9.** What is between the orbits of Mars and Jupiter?

a. Kuiper belt
b. asteroid belt
c. Oort cloud

____ **10.** A flash of light in the night sky is a(an)

a. meteor.
b. asteroid.
c. meteorite.

Completion

Read each word in the box. In each sentence below, fill in the correct word or words. Not all words will be used.

meteorite	sunspots	terrestrial	ring	ellipse	meteor

11. A(an) ______________________________ is an oval shape.

12. Areas of gas on the sun that are cooler than the gases around them are called ______________________________.

13. The inner planets are often called the ______________________________ planets.

14. A thin disk of small particles of ice and rock around a planet is a(an) ______________________________.

15. A meteoroid that passes through the atmosphere and hits Earth's surface is a(an) ______________________________.

True or False

If a statement is true, write true. *If it is false, write* false.

__________ **16.** A heliocentric model of the solar system is sun-centered.

__________ **17.** When you look at an image of the sun, you are looking at the chromosphere.

© Pearson Education, Inc., publishing as Pearson Prentice Hall. All rights reserved.

_________ 18. The four outer planets are often called gas giants.

_________ 19. A comet is a rocky object in the solar system that is too small to be a planet.

_________ 20. Life other than that on Earth would be called extraterrestrial life.

Using Science Skills

The picture shows a star and three planets that are orbiting the star. Use the picture to answer questions 21, 22, and 23.

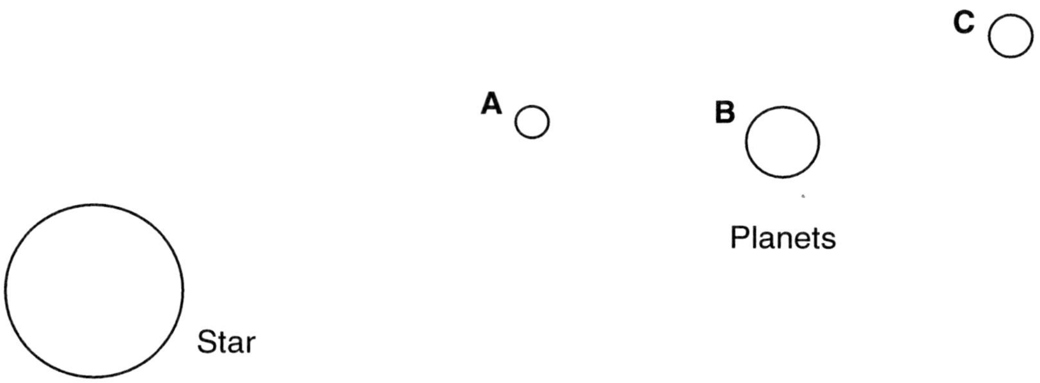

____ 21. **Interpreting Diagrams** Which planet is closest to the star?

a. Planet A

b. Planet B

c. Planet C

____ 22. **Inferring** Which planet will take the longest time to revolve around the star?

a. Planet A

b. Planet B

c. Planet C

____ 23. **Applying Concepts** Suppose the star is our sun and Planet B is Earth. In a heliocentric system, which would be true?

a. Planet B would revolve around Planet C.

b. The star would revolve around Planet B.

c. Planet B would revolve around the star.

© Pearson Education, Inc., publishing as Pearson Prentice Hall. All rights reserved.

Using Science Skills

The table below shows three characteristics of Jupiter's four largest moons. Use the table to answer questions 24 and 25.

Jupiter's Largest Moons			
Moon	**Diameter (km)**	**Average Distance from Jupiter (km)**	**Length of Revolution Around Jupiter (Earth days)**
Io	3,642	421,600	1.8
Europa	3,120	670,900	3.5
Ganymede	5,268	1,070,000	7.2
Callisto	4,800	1,883,000	16.7

____ **24. Interpreting Tables** Which of the four large moons of Jupiter is the largest?

a. Europa
b. Ganymede
c. Callisto

____ **25. Inferring** Why does Callisto take the longest to revolve around Jupiter?

a. Callisto has the greatest diameter of the four moons.
b. Callisto is the closest of the four moons to Jupiter.
c. Callisto is the farthest of the four moons away from Jupiter.

© Pearson Education, Inc., publishing as Pearson Prentice Hall. All rights reserved.

Stars, Galaxies, and the Universe

Multiple Choice

Write the letter of the correct answer on the line at the left.

____ **1.** The distance between the crest of one wave and the crest of the next wave is the wave's

a. radiation.

b. spectrum.

c. wavelength.

____ **2.** What is a building that has one or more telescopes?

a. observatory

b. radio telescope

c. planetesimal

____ **3.** Patterns of stars in the sky are called

a. star clusters.

b. constellations.

c. nebulas.

____ **4.** What do astronomers use to measure distances to nearby stars?

a. galaxies

b. parallax

c. sound waves

____ **5.** The earliest stage in a star's life is called a

a. quasar.

b. pulsar.

c. protostar.

____ **6.** How long a star "lives" depends on its

a. mass.

b. color.

c. temperature.

____ **7.** What is a huge group of single stars, star systems, star clusters, dust, and gas?

a. nebula

b. galaxy

c. black hole

© Pearson Education, Inc., publishing as Pearson Prentice Hall. All rights reserved.

____ 8. Young galaxies with giant black holes at their centers are
   a. quasars.
   b. pulsars.
   c. supernovas.

____ 9. According to Hubble's law, the farther away a galaxy is,
   a. the slower it is moving away from Earth.
   b. the sooner it will stop moving.
   c. the faster it is moving away from Earth.

____ 10. Astronomers think that in the future the universe will continue to
   a. shrink.
   b. expand.
   c. explode.

Completion

Read each word in the box. In each sentence below, fill in the correct word or words. Not all words will be used.

refracting	big bang	spectrograph
elliptical	reflecting	black hole

11. A(an) ______________________ telescope focuses light through lenses.
12. A device used by astronomers to discover the elements found in stars is a(an) ______________________.
13. An object with gravity so strong that no light can escape is a(an) ______________________.
14. A(an) ______________________ galaxy looks like a round or flattened ball.
15. The universe began with an explosion called the ______________________.

True or False

If a statement is true, write true. *If it is false, write* false.

________ 16. The light you can see is called visible light.

© Pearson Education, Inc., publishing as Pearson Prentice Hall. All rights reserved.

_________ **17.** A star's absolute brightness is the star's brightness as seen from Earth.

_________ **18.** Spinning neutron stars are called supernovas.

_________ **19.** The Milky Way is a spiral galaxy.

_________ **20.** Astronomers think that most of the universe is made of dark energy and dark matter.

Using Science Skills

The diagram below is used by astronomers to classify stars. Use the diagram to answer questions 21, 22, 23, 24, and 25.

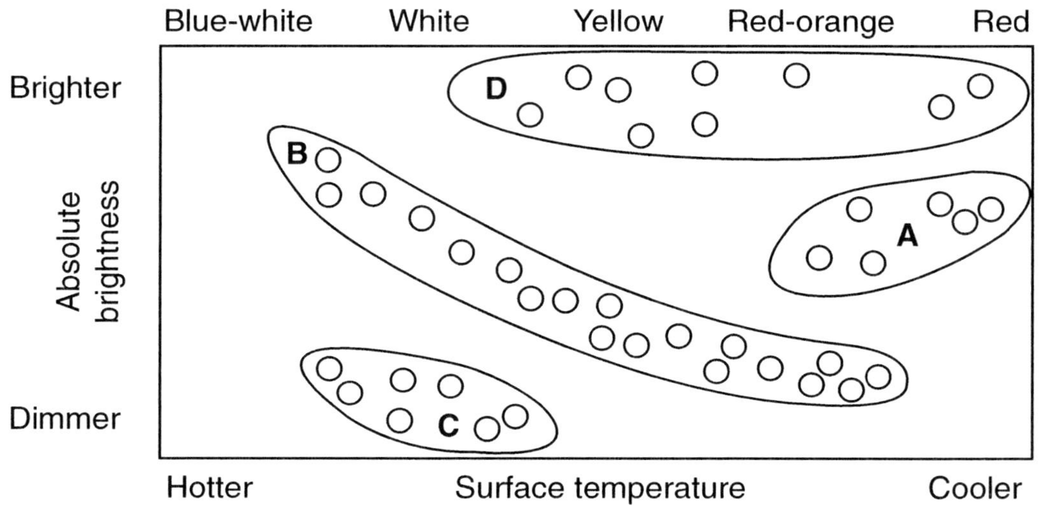

____ **21. Interpreting Graphs** What is the name for the kind of diagram shown in the picture?

a. H-R diagram

b. universal diagram

c. spectrograph

____ **22. Interpreting Graphs** Which group of stars in the diagram has the greatest absolute brightness?

a. group D

b. group B

c. group C

© Pearson Education, Inc., publishing as Pearson Prentice Hall. All rights reserved.

Using Science Skills (continued)

____ 23. **Interpreting Graphs** In which group of stars could a very bright red star or a very bright yellow star be found?

a. group B
b. group C
c. group D

____ 24. **Applying Concepts** Which group of stars represents white dwarfs?

a. group A
b. group B
c. group C

____ 25. **Applying Concepts** In which group of stars would a main-sequence star be found?

a. group A
b. group B
c. group C

© Pearson Education, Inc., publishing as Pearson Prentice Hall. All rights reserved.

Astronomy

Multiple Choice

Write the letter of the correct answer on the line at the left.

____ **1.** The study of the moon, stars, and other objects in space is called
- **a.** Hubble's law.
- **b.** astronomy.
- **c.** microgravity.

____ **2.** The force that attracts all objects toward each other is
- **a.** gravity.
- **b.** inertia.
- **c.** weight.

____ **3.** Who was the first person to walk on the moon?
- **a.** Yuri Gagarin
- **b.** Neil Armstrong
- **c.** John Glenn

____ **4.** What is a spacecraft that collects data in space but carries no humans?
- **a.** space shuttle
- **b.** space station
- **c.** space probe

____ **5.** A sun-centered model of the solar system is a
- **a.** heliocentric system.
- **b.** H-R diagram.
- **c.** geocentric system.

____ **6.** Which is the largest planet in the solar system?
- **a.** Earth
- **b.** Jupiter
- **c.** Saturn

____ **7.** A "dirty snowball" that orbits the sun is a(an)
- **a.** asteroid.
- **b.** comet.
- **c.** meteoroid.

____ **8.** The brightness of a star as seen from Earth is called
- **a.** parallax.
- **b.** absolute brightness.
- **c.** apparent brightness.

© Pearson Education, Inc., publishing as Pearson Prentice Hall. All rights reserved.

____ **9.** What unit do astronomers use to measure distances between stars?
  a. light-year
  b. scientific notation
  c. kilometer

____ **10.** The explosion of a supergiant star is called a
  a. pulsar.
  b. black hole.
  c. supernova.

Completion

Read each word in the box. In each sentence below, fill in one of the words. Not all words will be used.

reflecting	ellipse	galaxy
thrust	refracting	revolution

11. The movement of one object around another is a(an) ______________________.

12. The force that pushes a rocket forward is called ______________________.

13. An oval shape is called a(an) ______________________.

14. The Hubble Space Telescope is a(an) ______________________ telescope.

15. A huge group of single stars, star systems, star clusters, dust, and gas bound together by gravity is a(an) ______________________.

True or False

If a statement is true, write true. *If it is false, write* false.

________ **16.** Maria are mountains on the moon.

________ **17.** The space race was between the Soviet Union and China.

________ **18.** The sun gets its energy through the process of nuclear fusion.

© Pearson Education, Inc., publishing as Pearson Prentice Hall. All rights reserved.

_______ **19.** The planet Neptune is one of the gas giants.

_______ **20.** Astronomers think that the universe began with the big bang.

Using Science Skills

The picture below shows the moon orbiting around Earth. Use the picture to answer questions 21 and 22.

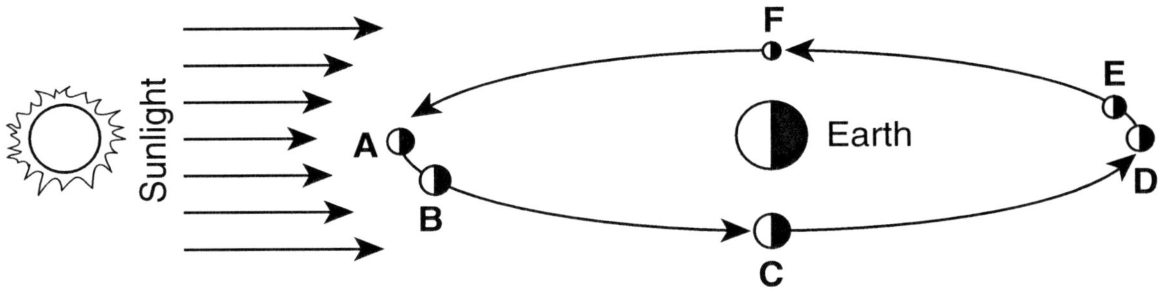

_____ **21. Interpreting Diagrams** At which position would the moon be when someone on Earth would see a full moon?

a. position A
b. position C
c. position D

_____ **22. Applying Concepts** What kind of tide would occur when the person sees a full moon?

a. spring tide
b. low tide
c. neap tide

© Pearson Education, Inc., publishing as Pearson Prentice Hall. All rights reserved.

Name ______________________ Date ______________ Class __________

Using Science Skills

The picture shows features of the sun's interior and atmosphere. Use the picture to answer questions 23, 24, and 25.

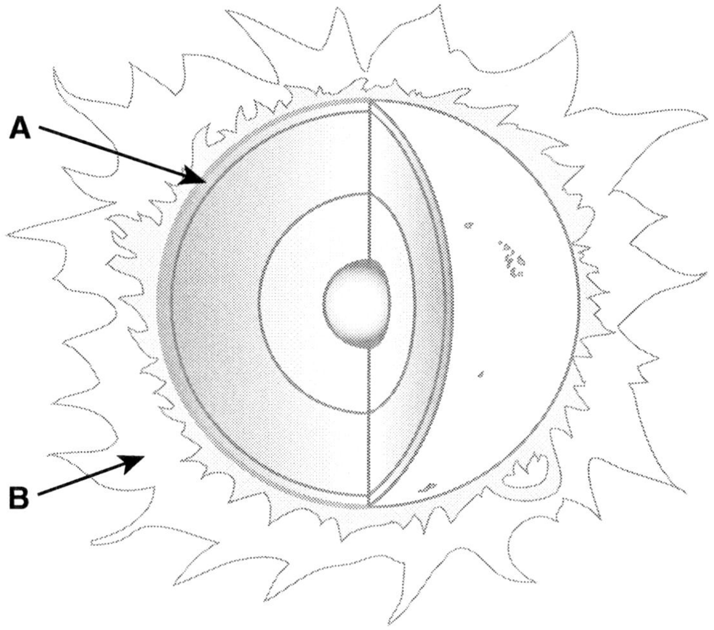

____ **23. Interpreting Diagrams** When you look at an image of the sun, you see the layer labeled A. What is the name of that layer of the sun?

a. core

b. photosphere

c. chromosphere

____ **24. Interpreting Diagrams** Which feature of the sun's atmosphere is labeled B?

a. prominence

b. solar wind

c. corona

____ **25. Applying Concepts** Which feature of the sun looks like a white halo around the sun during a total solar eclipse?

a. corona

b. chromosphere

c. solar flare

© Pearson Education, Inc., publishing as Pearson Prentice Hall. All rights reserved.

Chapter 1, Earth, Moon, and Sun

1. c
2. b
3. c
4. a
5. b
6. b
7. c
8. b
9. a
10. c
11. solstice
12. weight
13. neap tide
14. full moon
15. crater
16. false
17. true
18. false
19. true
20. true
21. c
22. a
23. c
24. b
25. a

Chapter 2, Exploring Space

1. b
2. b
3. a
4. a
5. b
6. c
7. b
8. b
9. a
10. b
11. velocity
12. satellite
13. rovers
14. vacuum
15. space spinoffs
16. true
17. false
18. false
19. true
20. true
21. b
22. a
23. a
24. c
25. b

Chapter 3, The Solar System

1. a
2. c
3. b
4. c
5. a
6. b
7. c
8. c
9. b
10. a
11. ellipse
12. sunspots
13. terrestrial
14. ring
15. meteorite
16. true
17. false
18. true
19. false
20. true
21. a
22. c
23. c
24. b
25. c

Chapter 4, Stars, Galaxies, and the Universe

1. c
2. a
3. b
4. b
5. c
6. a
7. b
8. a
9. c
10. b
11. refracting
12. spectrograph
13. black hole
14. elliptical
15. big bang
16. true
17. false
18. false
19. true

© Pearson Education, Inc., publishing as Pearson Prentice Hall. All rights reserved.

20. true
21. a
22. a
23. c
24. c
25. b

Book Test, Astronomy

1. b
2. a
3. b
4. c
5. a
6. b
7. b
8. c
9. a
10. c
11. revolution
12. thrust
13. ellipse
14. reflecting
15. galaxy
16. false
17. false
18. true
19. true
20. true
21. c
22. a
23. b
24. c
25. a

© Pearson Education, Inc., publishing as Pearson Prentice Hall. All rights reserved.